Born To Run

Written by Grace Niemiec
Illustrated by Sara Niemiec

The Virtuoso Press

Hey.

Hey, you.

Why did you flip the page?
I want to talk to you!

Now that I have your
attention, listen here!
I want to tell you my story.

I am a greyhound. I was born with strong legs and a skinny body that helped me run super fast.

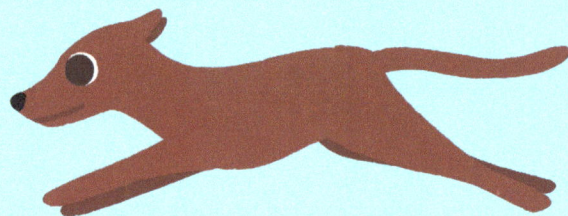

Even as a pup, I was faster than my friends.

There was no doubt about it! I was *born to run!*

I was quickly sent to a racetrack where I could prove my speed.

The humans at the
track knew I was the
best, so they named me
ACE.

Days at the track were tiring.

I ran.

I ate.

I slept.

But I was a champion!
There was nothing better
than being *born to run!*

I did get sad sometimes.
The crowds would leave.
The sun would set and
I would wonder if
something was missing.
Was racing all there was?

Maybe I would never know...

That changed one day. All the dogs were put onto trucks. I asked one what happened.

She said the track had been shut down. We didn't have to race anymore.

Shut down? Could my racing
days really be over?

As the truck started moving,
I felt scared.
If I was born to run,
what would I do now?

The truck drove for a while
and dropped us off with
some nice ladies.

At least I thought
they were nice.

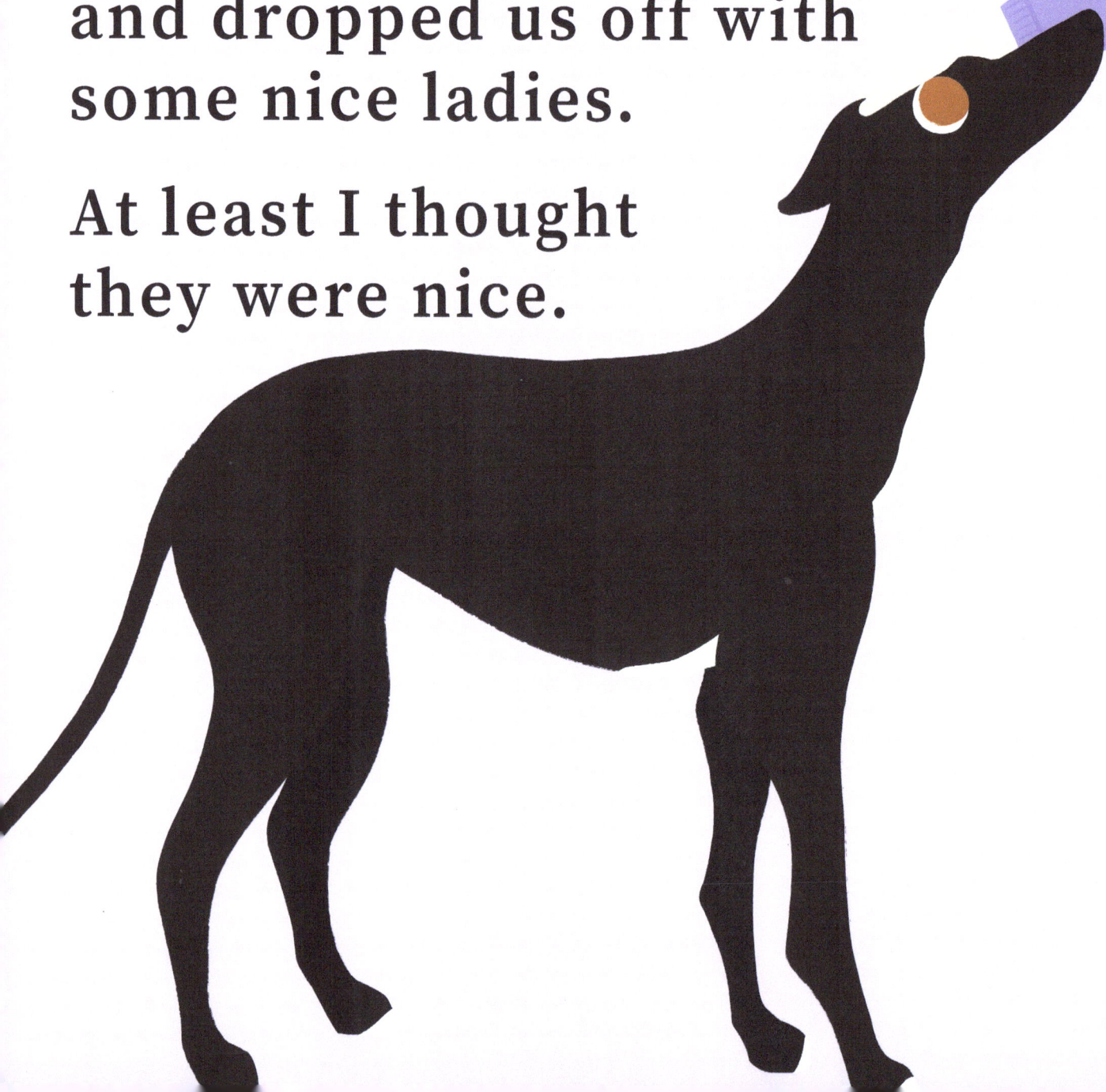

But they gave me a bath.

I don't like baths.

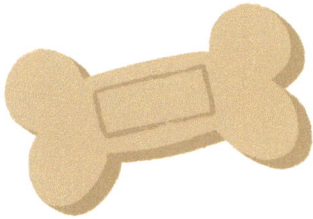

As I dried off in the sun, some people walked over.

Some pet me. One lady gave me a treat. They were all so nice, I almost forgot about the track.

A few treats and pets later,
a little girl came over to me.
She smelled like bananas!
I gave her a little lick to check.

She laughed and laughed.
I kept licking, wondering
where the bananas were hidden.

Two adults and a boy walked up behind her. They smiled.

"Is he the one, Lily?" a tall man asked. Lily said yes.

Soon, I was packed into a smaller truck. It was a little *too* small. But I could continue sniffing for bananas, so I didn't mind.

I learned that the boy's name was Ben and he did not smell like bananas. I still liked him because he scratched behind my ears.

The truck stopped
outside a cozy-looking
building. Lily told me this
was my new home.

"What is a *home*?" I thought to myself.

Home was nothing
like the track.
I was confused by
some things.

What are these?

What is this?

WHAT IS THAT?

Ace!

I was starting to wonder
if this "home" thing was a
good idea, but I heard Lily
call my name.

She put on my leash and
told me I had a surprise
behind the house.

"This is your yard," Lily said. She unclipped my leash and I sped off. I could run here!

Soon, Ben and the tall man joined me. They could almost run as fast as me!

It was just like the track, but with one tiny difference.

What had changed?
I thought all through
dinner, but was
distracted by the
scraps Ben dropped.

After my goodnight kiss from Lily,
I finally realized:
maybe I wasn't just
born to run...

...maybe I could be loved.

About Greyhounds

- They may be born to run, but greyhounds sleep an average of 18 hours per day!

- Most greyhounds race for 2-3 years before retiring. After that, they need homes.

- In 2020, the state of Florida closed all greyhound race tracks. Now, thousands of greyhounds need homes!

- Find more information on greyhound adoption at the link below.

★ Visit: www.greyhoundsonly.com ★

About Us

"Born To Run" is **Grace Niemiec**'s first book and is a part of her Girl Scout Gold Award project, the highest achievement a Girl Scout can receive. Grace is currently a high school senior in the Chicago suburbs and lives with her parents, sister, and three pets.

Sara Niemiec is an Elementary Education major at Augustana College, a sister to Grace Niemiec, and a gifted artist. "Born To Run" is Sara's first illustration project.

Ace Niemiec won plenty of races, but retired early due to a leg injury. Ace has found his forever home and is a happy dog that enjoys treats, forgetting how big he is, and playing with his stuffed hedgehog. He also really likes bananas.

Acknowledgements

Thank you to...

Brian Dolan and Virtuoso Press. Through mentorship and publishing, this book is better because of your help.

Mindy Young, for her continued efforts within Girl Scout Troop 41614. Her actions as a troop leader allowed for effective leadership within this project.

...as well as the following people:

Arnel Reynon • Nancy Genson • Tricia Goebel
Lisa Weistroffer • Todd Niemiec • Julie Niemiec

www.ingramcontent.com/pod-product-compliance
Lightning Source LLC
Chambersburg PA
CBHW051601190326
41458CB00030B/6498